I0500359

Grandma's Quilts

Fun with Fractions

Kathleen L. Stone

Enjoy these other books by Kathleen L. Stone

Penguin Place Value
A Math Adventure

Number Line Fun
Solving Number Mysteries

Riley the Robot
An Input/Output Machine

Mason the Magician
Hundreds Chart Addition

Katelyn's Fair Share Picnic
More Math Fun

Money Tree Mysteries
Adventures with Quarters

Alien Even and Alien Odd
A Math Space Adventure

Kenley's Line Plot Graph
Another Math Adventure

Matthew's Sunshine Bakery
Multiplication Arrays

Firefighter Gary's Fire Safety Rules

From My Quilted Heart to Yours (Book 1)
Heart Warming Quilts and Heart Healthy Recipes for Your Loved Ones

From My Quilted Heart to Yours (Book 2)
Quilts and Blocks from the Children's Book, Grandma's Quilts

Copyright © 2015 Kathleen L. Stone

All rights reserved.

ISBN–13: 978-1515196716
ISBN-10: 1515196712

Dedication

To my beautiful grandchildren … you have brought much happiness and love to my life.

My grandma loves to quilt
But little did I know
That she would help me with my math
As she taught me how to sew.

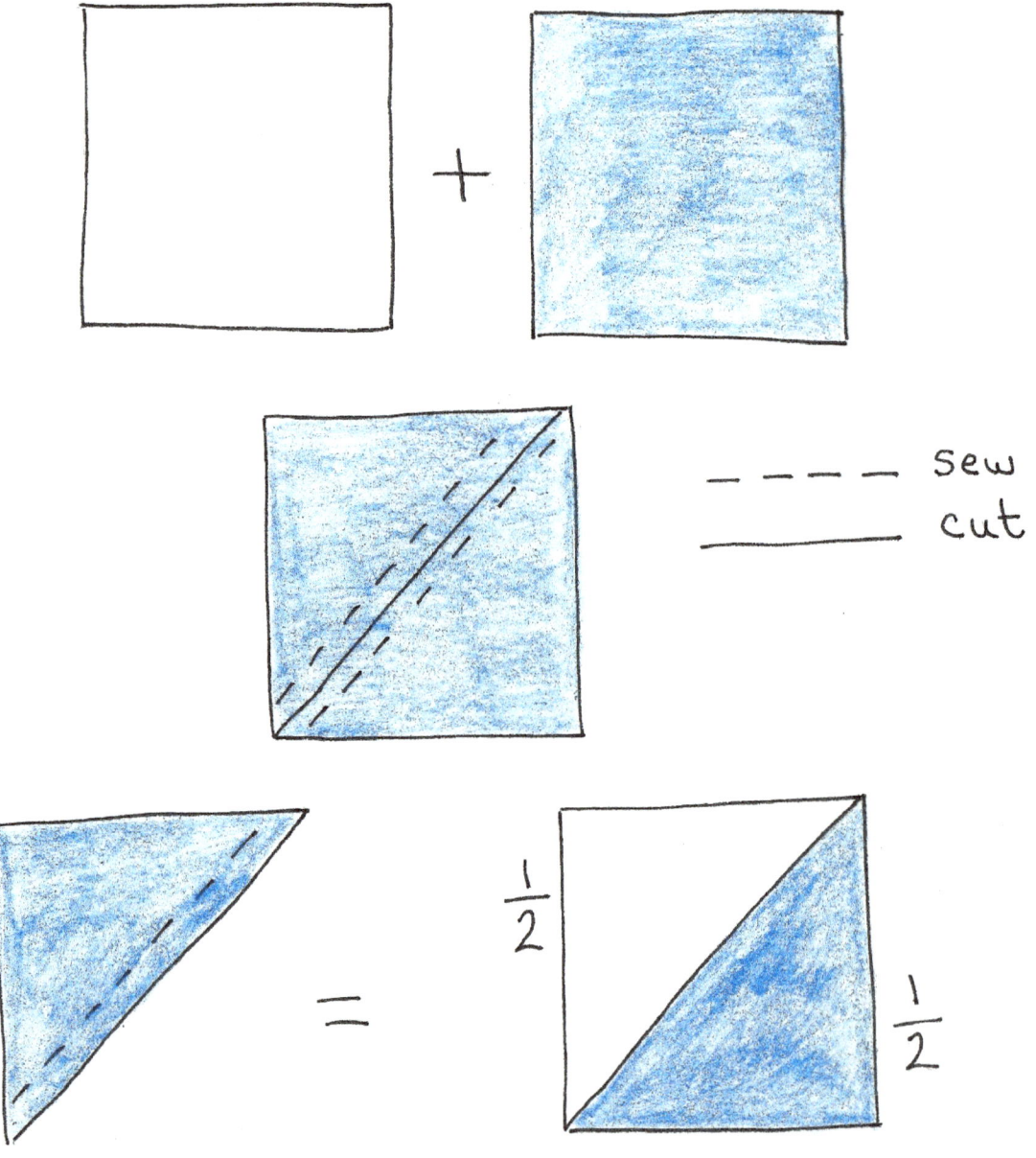

We took two squares of fabric.
Sewed them together, then cut them in two.
Just look at the block we made.
½ is white and ½ is blue.

Using extra pieces of our fabric
We sewed and cut some more.
When I counted the squares in this
block
I could see that there were four.

Then Grandma asked me to tell her
What fraction of this block was white?
When I told her it was *one half*
She said that I was right!

We used the rest of our fabric
Measured, sewed, and cut pieces apart.
And when we were all finished
We had made this patchwork heart!

Here's another block we made.
Please tell me what you think.
What fraction describes
How much of this block is pink?

If you said *one half*
You are absolutely right.
One half of it is pink
And *one half* of it is white.

Grandma and I made thirty-two blocks.
We made them pretty quick.
Then donated the quilt to our hospital
For a little girl who was sick.

$\frac{1}{4}$

$\frac{1}{4}$

$\frac{1}{4}$

$\frac{1}{4}$

But if we used all different fabrics
Then what fraction would you see?
Each fabric shows *one fourth*
And are as pretty as can be.

So then I told my Grandma
That I'd really like to see
More blocks that show *fourths*.
Would she please help me?

Grandma said it would be easy
We had lots of scraps to use.
The only thing to do now
Was decide which block to choose.

I chose the *Flying Geese* block.
What fraction do you see?
Each goose is a different fabric.
Looks like each goose is *one fourth* to
me.

Here's another quilt we made.
In the center we appliquéd birds.
Look carefully at each block.
Do you see they're divided in *thirds*?

How can we tell it's divided in *thirds*?
I'll give you a little hint.
Each block has *three* strips of fabric.
But each strip is a different print.

This quilt used the same block.
How many strips do you see?
Let's count them together.
There's *one, two, three.*

Another block with *three* strips
Put together in a new design.
But they still show *one third.*
We're giving this quilt to a friend of
mine.

Patchwork Hearts

T-Shirt Quilt

Geese and Goslings

Traffic Jam

Picnics-n-Pink
Lemonade

Bird Nest ...
Home Tweet Home

Cotton Candy Dreams

Panda Puzzle

I see fractions!

Grandma and I made lots of quilts.
Some we kept. Others we gave away.
But I know that I will never
Look at quilts in quite the same way.

What
fractions do
you see?

Every block that I make.
Will fill me with satisfaction
Because I'm really getting better
At learning all my fractions.

Fractions … Parts of a Whole

Fractions are formed when a whole is divided into **equal parts.** Children need time to explore and manipulate objects to truly gain an understanding about fractions. Using objects that children are familiar with will aid in their understanding. Quilt blocks make a great visual for children to use to identify fractions. *From My Quilted Heart to Yours (Book 2)* provides directions on the blocks and quilts used in *Grandma's Quilts*.

Enrichment Activities

Sharing a Cookie
Materials needed:

paper plates (4 for each child)
crayons
scissors

Give each child four paper plates. Have them decorate each plate as their favorite cookie. Using one of their cookies, ask how much of the cookie they would get if they were the only one having a cookie (*the whole cookie*). How would that change if they had to share that cookie with a friend? Have them brainstorm how they would share the cookie so that each of them got the same amount. Eventually help them discover that if the cookie were cut in *half*, they would each get a piece of the cookie (*fold the paper plate in half and cut it into two pieces*). Explain that each of them would get *one half* of the cookie. Repeat with the other "cookies" to show *thirds* and *fourths*.

Playdough Fractions
Materials needed:

different colored playdough
plastic knife

Have each child make simple shapes (circles, squares, etc.) with their playdough and use plastic knives to divide each share into equal parts. Use different colored playdough to help identify fractions (see illustration below). One fourth of the square is blue, one fourth is red, etc.

Cracked Egg Hunt

Materials needed:

plastic eggs
pictures of fractions
fraction labels (that match the pictures)

Tape a fraction picture on one end of the egg and the matching fraction label on the other end (to make this activity more challenging, make sure the colors of each egg half do **not** match). Open the eggs and hide the "cracked" eggs around the house. Have the children go on a hunt, putting their eggs together so that the fraction pictures and labels match.

Ice Cream Cone Fractions

Materials needed:

a variety of white (vanilla), pink (strawberry), and brown (chocolate) paper ice cream scoops
several paper ice cream cones

Have children make a variety of ice cream cones using only three scoops per cone. Have them identify the fractions of vanilla, strawberry, and chocolate they have on each cone (see examples below).

1/3 is chocolate
2/3 is strawberry

1/3 is vanilla
1/3 is chocolate
1/3 is strawberry

ABOUT THE AUTHOR

Kathleen Stone is a National Board Certified educator and is currently teaching second grade. *Grandma's Quilts* is her eleventh children book published. She loves spending time with her family. She and her husband Gary live in the Olympia area but enjoy traveling across the United States. When not teaching, Kathleen can often be found quilting, sitting by the lake reading, or exploring new parks with her grandchildren!

Math is all around us
No matter where you turn
Open your mind to the wonders of math
And all that you can learn

www.ingramcontent.com/pod-product-compliance
Lightning Source LLC
Chambersburg PA
CBHW050406180526
45159CB00005B/2168